OPERATION HAYSTACK

Frank Herbert

Illustrator : H. R. van Dongen

Operation Haystack

Table of Contents
Operation Haystack

Operation Haystack

It's hard to ferret out a gang of fanatics; it would, obviously, be even harder to spot a genetic line of dedicated men. But the problem Orne had was one step tougher than that!

When the Investigation & Adjustment scout cruiser landed on Marak it carried a man the doctors had no hope of saving. He was alive only because he was in a womblike creche pod that had taken over most of his vital functions.

The man's name was Lewis Orne. He had been a blocky, heavy–muscled redhead with slightly off–center features and the hard flesh of a heavy planet native. Even in the placid repose of near death there was something clownish about his appearance. His burned, ungent–covered face looked made up for some bizarre show.

Marak is the League capital, and the I–A medical center there is probably the best in the galaxy, but it accepted the creche pod and Orne more as a curiosity than anything else. The man had lost one eye, three fingers of his left hand and part of his hair, suffered a broken jaw and various internal injuries. He had been in terminal shock for more than ninety hours.

Umbo Stetson, Orne's section chief, went back into his cruiser's "office" after a hospital flitter took pod and patient. There was an added droop to Stetson's shoulders that accentuated his usual slouching stance. His overlarge features were drawn into ridges of sorrow. A general straggling, trampish look about him was not helped by patched blue fatigues.

The doctor's words still rang in Stetson's ears: "This patient's vital tone is too low to permit operative replacement of damaged organs. He'll live for a while because of the pod, but—" And the doctor had shrugged.

Stetson slumped into his desk chair, looked out the open port beside him. Some four hundred meters below, the scurrying beetlelike activity of the I–A's main field sent up discordant roaring and clattering. Two rows of other scout cruisers were parked in line with Stetson's port—gleaming red and black needles. He stared at them without really seeing them.

It always happens on some "routine" assignment, he thought. *Nothing but a slight suspicion about Heleb: the fact that only women held high office. One simple, unexplained fact ... and I lose my best agent!*

He sighed, turned to his desk, began composing the report:

"The militant core on the Planet Heleb has been eliminated. Occupation force on the ground. No further danger to Galactic peace expected from this source. Reason for operation: Rediscovery & Re–education—*after two years on the planet*—failed to detect signs of militancy. The major indications were: 1) a ruling caste restricted to women, and 2) disparity between numbers of males and females *far* beyond the Lutig norm! Senior Field Agent Lewis Orne found that the ruling caste was controlling the sex of offspring at conception (see attached details), and had raised a male slave army to maintain its rule. The R&R agent had been drained of information, then killed. Arms constructed on the basis of that information caused critical injuries to Senior Field Agent Orne. He is not expected to live. I am hereby urging that he receive the Galaxy Medal, and that his name be added to the Roll of Honor."

Stetson pushed the page aside. That was enough for ComGO, who never read anything but the first page anyway. Details were for his aides to chew and digest. They could wait. Stetson punched his desk callbox for Orne's service record, set himself to the task he most detested: notifying next of kin. He read, pursing his lips:

"Home Planet: Chargon. Notify in case of accident or death: Mrs. Victoria Orne, mother."

He leafed through the pages, reluctant to send the hated message. Orne had enlisted in the Marak Marines at age seventeen—a runaway from home—and his mother had given post–enlistment consent. Two years later: scholarship transfer to Uni–Galacta, the R&R school here on Marak. Five years of school and one R&R field assignment under his belt, and he had been drafted into the I–A for brilliant detection of militancy on Hammel. And two years later—*kaput!*

Abruptly, Stetson hurled the service record at the gray metal wall across from him; then he got up, brought the record back to his desk, smoothing the pages. There were tears in his eyes. He flipped a switch on his desk, dictated the notification to Central Secretarial, ordered it sent out priority. Then he went groundside and got drunk on Hochar Brandy, Orne's favorite drink.

* * * * *

The next morning there was a reply from Chargon: "Lewis Orne's mother too ill to travel. Sisters being notified. Please ask Mrs. Ipscott Bullone of Marak, wife of the High Commissioner, to take over for family." It was signed: "Madrena Orne Standish, sister."

With some misgivings, Stetson called the residence of Ipscott Bullone, leader of the majority party in the Marak Assembly. Mrs. Bullone took the call with blank screen. There was a sound of running water in the background. Stetson stared at the grayness swimming in his desk visor. He always disliked a blank screen. A baritone husk of a voice slid: "This is Polly Bullone."

Stetson introduced himself, relayed the Chargon message.

"Victoria's boy dying? Here? Oh, the poor thing! And Madrena's back on Chargon ... the election. Oh, yes, of course. I'll get right over to the hospital!"

Stetson signed off, broke the contact.

The High Commissioner's wife yet! he thought. Then, because he had to do it, he walled off his sorrow, got to work.

At the medical center, the oval creche containing Orne hung from ceiling hooks in a private room. There were humming sounds in the dim, watery greenness of the room, rhythmic chuggings, sighings. Occasionally, a door opened almost soundlessly, and a white–clad figure would check the graph tapes on the creche's meters.

Orne was lingering. He became the major conversation piece at the internes' coffee breaks: "That agent who was hurt on Heleb, he's still with us. Man, they must build those guys different from the rest of us!...Yeah! Understand he's got only about an eighth of his insides ... liver, kidneys, stomach—all gone.... Lay you odds he doesn't last out the month.... Look what old sure–thing McTavish wants to bet on!"

On the morning of his eighty–eighth day in the creche, the day nurse came into Orne's room, lifted the inspection hood, looked down at him. The day nurse was a tall, lean–faced professional who had learned to meet miracles and failures with equal lack of expression. However, this routine with the dying I–A operative had lulled her into a state of psychological unpreparedness. *Any day now, poor guy*, she thought. And she gasped as she opened his sole remaining eye, said:

"Did they clobber those dames on Heleb?"

"Yes, sir!" she blurted. "They really did, sir!"

"Good!"

Orne closed his eye. His breathing deepened.

The nurse rang frantically for the doctors.

It had been an indeterminate period in a blank fog for Orne, then a time of pain and the gradual realization that he was in a creche. Had to be. He could remember his sudden exposure on Heleb, the explosion—then nothing. Good old creche. It made him feel safe now, shielded from all danger.

Orne began to show minute but steady signs of improvement. In another month, the doctors ventured an intestinal graft that gave him a new spurt of energy. Two months later, they replaced missing eye and fingers, restored his scalp line, worked artistic surgery on his burn scars.

Fourteen months, eleven days, five hours and two minutes after he had been picked up "as good as dead," Orne walked out of the hospital under his own power, accompanied by a strangely silent Umbo Stetson.

Under the dark blue I–A field cape, Orne's coverall uniform fitted his once muscular frame like a deflated bag. But the pixie light had returned to his eyes—even to the eye he had received from a nameless and long dead donor. Except for the loss of weight, he looked to be the same Lewis Orne. If he was different—beyond the "spare parts"—it was something he only suspected, something that made the idea, "twice–born," not a joke.

* * * * *

Outside the hospital, clouds obscured Marak's green sun. It was midmorning. A cold spring wind bent the pile lawn, tugged fitfully at the border plantings of exotic flowers around the hospital's landing pad.

Orne paused on the steps above the pad, breathed deeply of the chill air. "Beautiful day," he said.

Stetson reached out a hand to help Orne down the steps, hesitated, put the hand back in his pocket. Beneath the section chief's look of weary superciliousness there was a note of anxiety. His big features were set in a frown. The drooping eyelids failed to conceal a sharp, measuring stare.

Orne glanced at the sky to the southwest. "The flitter ought to be here any minute." A gust of wind tugged at his cape. He staggered, caught his balance. "I *feel* good."

"You look like something left over from a funeral," growled Stetson.

"Sure—my funeral," said Orne. He grinned. "Anyway, I was getting tired of that walk–around–type morgue. All my nurses were married."

"I'd almost stake my life that I could trust you," muttered Stetson.

Orne looked at him. "No, no, Stet ... stake *my* life. I'm used to it."

Stetson shook his head. "No, dammit! I trust you, but you deserve a peaceful convalescence. We've no right to saddle you with—"

"Stet?" Orne's voice was low, amused.

"Huh?" Stetson looked up.

"Let's save the noble act for someone who doesn't know you," said Orne. "You've a job for me. O.K. You've made the gesture for your conscience."

Stetson produced a wolfish grin. "All right. So we're desperate, and we haven't much time. In a nutshell, since you're going to be a house guest at the Bullones'—we suspect Ipscott Bullone of being the head of a conspiracy to take over the government."

"What do you mean—*take over the government*?" demanded Orne. "The Galactic High Commissioner *is* the government—subject to the Constitution and the Assemblymen who elected him."

"We've a situation that could explode into another Rim War, and we think he's at the heart of it," said Stetson. "We've eighty-one touchy planets, all of them old-line steadies that have been in the League for years. And on every one of them we have reason to believe there's a clan of traitors sworn to overthrow the League. Even on your home planet—Chargon."

"You want me to go home for my convalescence?" asked Orne. "Haven't been there since I was seventeen. I'm not sure that—"

"No, dammit! We want you as the Bullones' house guest! And speaking of that, would you mind explaining how they were chosen to ride herd on you?"

"There's an odd thing," said Orne. "All those gags in the I–A about old Upshook Ipscott Bullone ... and then I find that his wife went to school with my mother."

"Have you met Himself?"

"He brought his wife to the hospital a couple of times."

Again, Stetson looked to the southwest, then back to Orne. A pensive look came over his face. "Every schoolkid knows how the Nathians and the Marakian League fought it out in the Rim War—how the old civilization fell apart—and it all seems kind of distant," he said.

"Five hundred standard years," said Orne.

"And maybe no farther away than yesterday," murmured Stetson. He cleared his throat.

* * * * *

And Orne wondered why Stetson was moving so cautiously. *Something deep troubling him.* A sudden thought struck Orne. He said: "You spoke of trust. Has this conspiracy involved the I–A?"

"We think so," said Stetson. "About a year ago, an R&R archeological team was nosing around some ruins on Dabih. The place was all but vitrified in the Rim War, but a whole bank of records from a Nathian outpost escaped." He glanced sidelong at Orne. "The Rah&Rah boys couldn't make sense out of the records. No surprise. They called in an I-A crypt–analyst. He broke a complicated substitution cipher. When the stuff started making sense he pushed the panic button."

"For something the Nathians wrote five hundred years ago?"

Stetson's drooping eyelids lifted. There was a cold quality to his stare. "This was a routing station for key Nathian families," he said. "Trained refugees. An old dodge ... been used as long as there've been—"

"But five hundred *years*, Stet!"

"I don't care if it was five *thousand* years!" barked Stetson. "We've intercepted some scraps since then that were written in the *same* code. The bland confidence of *that*! Wouldn't that gall you?" He shook his head. "And every scrap we've intercepted deals with the coming elections."

"But the election's only a couple of days off!" protested Orne.

Stetson glanced at his wristchrono. "Forty–two hours to be exact," he said. "Some deadline!"

"Any names in these old records?" asked Orne.

Stetson nodded. "Names of planets, yes. People, no. Some code names, but no cover names. Code name on Chargon was *Winner*. That ring any bells with you?"

Orne shook his head. "No. What's the code name here?"

"The Head," said Stetson. "But what good does that do us? They're sure to've changed those by now."

"They didn't change their communications code," said Orne.

"No ... they didn't."

"We must have something on them, some leads," said Orne. He felt that Stetson was holding back something vital.

"Sure," said Stetson. "We have history books. They say the Nathians were top drawer in political mechanics. We know for a fact they chose landing sites for their *refugees* with diabolical care. Each family was told to dig in, grow up with the adopted culture, develop the weak spots, build an underground, train their descendants to take over. They set out to bore from within, to make victory out of defeat. The Nathians were long on patience. They came originally from nomad stock on Nathia II. Their mythology calls them Arbs or Ayrbs. Go review your seventh grade history. You'll know almost as much as we do!"

"Like looking for the traditional needle in the haystack," muttered Orne. "How come you suspect High Commissioner Upshook?"

Stetson wet his lips with his tongue. "One of the Bullones' seven daughters is currently at home," he said. "Name's Diana. A field leader in the I–A women. One of the Nathian code messages we intercepted had her name as addressee."

"Who sent the message?" asked Orne. "What was it all about?"

Stetson coughed. "You know, Lew, we cross–check everything. This message was signed M.O.S. The only M.O.S. that came out of the comparison was on a routine next–of–kin reply. We followed it down to the original copy, and the handwriting checked. Name of Madrena Orne Standish."

"Maddie?" Orne froze, turned slowly to face Stetson. "So that's what's troubling you!"

"We know you haven't been home since you were seventeen," said Stetson. "Your record with us is clean. The question is—"

"Permit me," said Orne. "The question is: Will I turn in my own sister if it falls that way?"

Stetson remained silent, staring at him.

"O.K.," said Orne. "My job is seeing that we don't have another Rim War. Just answer me one question: How's Maddie mixed up in this? My family isn't one of these traitor clans."

"This whole thing is all tangled up with politics," said Stetson. "We think it's because of her husband."

"Ahhhh, the member for Chargon," said Orne. "I've never met him." He looked to the southwest where a flitter was growing larger as it approached. "Who's my cover contact?"

"That mini–transceiver we planted in your neck for the Gienah job," said Stetson. "It's still there and functioning. Anything happens around you, we hear it."

Orne touched the subvocal stud at his neck, moved his speaking muscles without opening his mouth. A surf–hissing voice filled the matching transceiver in Stetson's neck:

"You pay attention while I'm making a play for this Diana Bullone, you hear? Then you'll know how an expert works."

"Don't get so interested in your work that you forget why you're out there," growled Stetson.

* * * * *

Mrs. Bullone was a fat little mouse of a woman. She stood almost in the center of the guest room of her home, hands clasped across the paunch of a long, dull silver gown. She had demure gray eyes, grandmotherly gray hair combed straight back in a jeweled net—and that shocking baritone husk of a voice issuing from a small mouth. Her figure sloped out from several chins to a matronly bosom, then dropped straight like a barrel. The top of her head came just above Orne's dress epaulets.

"We want you to feel at home here, Lewis," she husked. "You're to consider yourself one of the family."

Orne looked around at the Bullone guest room: low key furnishings with an old-fashioned selectacol for change of decor. A polawindow looked out onto an oval swimming pool, the glass muted to dark blue. It gave the outside a moonlight appearance. There was a contour bed against one wall, several built-ins, and a door partly open to reveal bathroom tiles. Everything traditional and comfortable.

"I already *do* feel at home," he said. "You know, your house is very like our place on Chargon. I was surprised when I saw it from the air. Except for the setting, it looks almost identical."

"I guess your mother and I shared ideas when we were in school," said Polly. "We were *very* close friends."

"You must've been to do all this for me," said Orne. "I don't know how I'm ever going to—"

"Ah! Here we are!" A deep masculine voice boomed from the open door behind Orne. He turned, saw Ipscott Bullone, High Commissioner of the Marakian League. Bullone was tall, had a face of harsh angles and deep lines, dark eyes under heavy brows, black hair trained in receding waves. There was a look of ungainly clumsiness about him.

He doesn't strike me as the dictator type, thought Orne. *But that's obviously what Stet suspects.*

"Glad you made it out all right, son," boomed Bullone. He advanced into the room, glanced around. "Hope everything's to your taste here."

"Lewis was just telling me that our place is very like his mother's home on Chargon," said Polly.

"It's old fashioned, but we like it," said Bullone. "Just a great big tetragon on a central pivot. We can turn any room we want to the sun, the shade or the breeze, but we usually leave the main salon pointing northeast. View of the capital, you know."

"We have a sea breeze on Chargon that we treat the same way," said Orne.

"I'm sure Lewis would like to be left alone for a while now," said Polly. "This is his first day out of the hospital. We mustn't tire him." She crossed to the polawindow, adjusted it to neutral gray, turned the selectacol, and the room's color dominance shifted to green. "There, that's more restful," she said. "Now, if there's anything you need you just ring the bell there by your bed. The autobutle will know where to find us."

The Bullones left, and Orne crossed to the window, looked out at the pool. The young woman hadn't come back. When the chauffeur–driven limousine flitter had dropped down to the house's landing pad, Orne had seen a parasol and sunhat nodding to each other on the blue tiles beside the pool. The parasol had shielded Polly Bullone. The sunhat had been worn by a shapely young woman in swimming tights, who had rushed off into the house.

She was no taller than Polly, but slender and with golden red hair caught under the sunhat in a swimmer's chignon. She was not beautiful—face too narrow with suggestions of Bullone's cragginess, and the eyes overlarge. But her mouth was full–lipped, chin strong, and there had been an air of exquisite assurance about her. The total effect had been one of striking elegance—extremely feminine.

Orne looked beyond the pool: wooded hills and, dimly on the horizon, a broken line of mountains. The Bullones lived in expensive isolation. Around them stretched miles of wilderness, rugged with planned neglect.

Time to report in, he thought. Orne pressed the neck stud on his transceiver, got Stetson, told him what had happened to this point.

"All right," said Stetson. "Go find the daughter. She fits the description of the gal you saw by the pool."

"That's what I was hoping," said Orne.

He changed into light–blue fatigues, went to the door of his room, let himself out into a hall. A glance at his wristchrono showed that it was shortly before noon—time for a bit of scouting before they called lunch. He knew from his brief tour of the house and its similarity to the home of his childhood that the hall let into the main living salon. The public rooms and men's quarters were in the outside ring. Secluded family apartments and women's quarters occupied the inner section.

* * * * *

Orne made his way to the salon. It was long, built around two sections of the tetragon, and with low divans beneath the view windows. The floor was thick pile rugs pushed one against another in a crazy patchwork of reds and browns. At the far end of the room, someone in blue fatigues like his own was bent over a stand of some sort. The figure straightened at the same time a tinkle of music filled the room. He recognized the red–gold hair of the young woman he had seen beside the pool. She was wielding two mallets to play a stringed instrument that lay on its side supported by a carved–wood stand.

He moved up behind her, his footsteps muffled by the carpeting. The music had a curious rhythm that suggested figures dancing wildly around firelight. She struck a final chord, muted the strings.

"That makes me homesick," said Orne.

"Oh!" She whirled, gasped, then smiled. "You startled me. I thought I was alone."

"Sorry. I was enjoying the music."

"I'm Diana Bullone," she said. "You're Mr. Orne."

"Lew to all of the Bullone family, I hope," he said.

"Of course ... Lew." She gestured at the musical instrument. "This is very old. Most find its music ... well, rather weird. It's been handed down for generations in mother's family."

"The kaithra," said Orne. "My sisters play it. Been a long time since I've heard one."

"Oh, of course," she said. "Your mother's—" She stopped, looked confused. "I've got to get used to the fact that you're.... I mean that we have a strange man around the house who isn't *exactly* strange."

Orne grinned. In spite of the blue I–A fatigues and a rather severe pulled–back hairdo, this was a handsome woman. He found himself liking her, and this caused him a feeling near self–loathing. She was a suspect. He couldn't afford to like her. But the Bullones were being so decent, taking him in like this. And how was their hospitality being repaid? By spying and prying. Yet, his first loyalty belonged to the I–A, to the peace it represented.

He said rather lamely: "I hope you get over the feeling that I'm strange."

"I'm over it already," she said. She linked arms with him, said: "If you feel up to it, I'll take you on the deluxe guided tour."

By nightfall, Orne was in a state of confusion. He had found Diana fascinating, and yet the most comfortable woman to be around that he had ever met. She liked swimming, *paloika* hunting, *ditar* apples— She had a "poo–poo" attitude toward the older generation that she said she'd never before revealed to anyone. They had laughed like fools over utter nonsense.

Orne went back to his room to change for dinner, stopped before the polawindow. The quick darkness of these low latitudes had pulled an ebon blanket over the landscape. There was city–glow off to the left, and an orange halo to the peaks where Marak's three moons would rise. *Am I falling in love with this woman?* he asked himself. He felt like calling Stetson, not to report but just to talk the situation out. And this made him acutely aware that Stetson or an aide had heard everything said between them that afternoon.

* * * * *

The autobutle called dinner. Orne changed hurriedly into a fresh lounge uniform, found his way to the small salon across the house. The Bullones already were seated around an old-fashioned bubble-slot table set with real candles, golden *shardi* service. Two of Marak's moons could be seen out the window climbing swiftly over the peaks.

"You turned the house," said Orne.

"We like the moonrise," said Polly. "It seems more romantic, don't you think?" She glanced at Diana.

Diana looked down at her plate. She was wearing a low-cut gown of *firemesh* that set off her red hair. A single strand of *Reinach* pearls gleamed at her throat.

Orne sat down in the vacant seat opposite her. *What a handsome woman!* he thought.

Polly, on Orne's right, looked younger and softer in a green stola gown that hazed her barrel contours. Bullone, across from her, wore black lounging shorts and knee-length *kubi* jacket of golden pearl cloth. Everything about the people and setting reeked of wealth, power. For a moment, Orne saw that Stetson's suspicions could have basis in fact. Bullone might go to any lengths to maintain this luxury.

Orne's entrance had interrupted an argument between Polly and her husband. They welcomed him, went right on without inhibition. Rather than embarrassing him, this made him feel more at home, more accepted.

"But I'm not running for office this time," said Bullone patiently. "Why do we have to clutter up the evening with that many people just to—"

"Our election night parties are traditional," said Polly.

"Well, I'd just like to relax quietly at home tomorrow," he said. "Take it easy with just the family here and not have to—"

"It's not like it was a *big* party," said Polly. "I've kept the list to fifty."

Diana straightened, said: "This is an important election Daddy! How could you *possibly* relax? There're seventy–three seats in question ... the whole balance. If things go wrong in just the Alkes sector ... why ...you could be sent back to the floor. You'd lose your job as ... why ...someone else could take over as—"

"Welcome to the job," said Bullone. "It's a headache." He grinned at Orne. "Sorry to burden you with this, m'boy, but the women of this family run me ragged. I guess from what I hear that you've had a pretty busy day, too." He smiled paternally at Diana. "And your first day out of the hospital."

"She sets quite a pace, but I've enjoyed it," said Orne.

"We're taking the small flitter for a tour of the wilderness area tomorrow," said Diana. "Lew can relax all the way. I'll do the driving."

"Be sure you're back in plenty of time for the party," said Polly. "Can't have—" She broke off at a low bell from the alcove behind her. "That'll be for me. Excuse me, please ... no, don't get up."

* * * * *

Orne bent to his dinner as it came out of the bubble slot beside his plate: meat in an exotic sauce, *Sirik* champagne, *paloika au semil* ...more luxury.

Presently, Polly returned, resumed her seat.

"Anything important?" asked Bullone.

"Only a cancellation for tomorrow night. Professor Wingard is ill."

"I'd just as soon it was cancelled down to the four of us," said Bullone.

Unless this is a pose, this doesn't sound like a man who wants to grab more power, thought Orne.

"Scottie, you should take more pride in your office!" snapped Polly. "You're an important man."

"If it weren't for you, I'd be a nobody and prefer it," said Bullone. He grinned at Orne. "I'm a political idiot compared to my wife. Never saw anyone who could call the turn like she does. Runs in her family. Her mother was the same way."

Orne stared at him, fork raised from plate and motionless. A sudden idea had exploded in his mind.

"You must know something of this life, Lewis," said Bullone. "Your father was member for Chargon once, wasn't he?"

"Yes," murmured Orne. "But that was before I was born. He died in office." He shook his head, thought: *It couldn't be ... but—*

"Do you feel all right, Lew?" asked Diana. "You're suddenly so pale."

"Just tired," said Orne. "Guess I'm not used to so much activity."

"And I've been a beast keeping you so busy today," she said.

"Don't you stand on ceremony here, son," said Polly. She looked concerned. "You've been very sick, and we understand. If you're tired, you go right on into bed."

Orne glanced around the table, met anxious attention in each face. He pushed his chair back, said: "Well, if you really don't mind—"

"Mind!" barked Polly. "You scoot along now!"

"See you in the morning, Lew," said Diana.

He nodded, turned away, thinking: *What a handsome woman!* As he started down the hall, he heard Bullone say to Diana: "Di, perhaps you'd better not take that boy out tomorrow. After all, he *is* supposed to be here for a rest." Her answer was lost as Orne entered the hall, closed the door.

In the privacy of his room, Orne pressed the transceiver stud at his neck, said: *"Stet?"*

A voice hissed in his ears: *"This is Mr. Stetson's relief. Orne, isn't it?"*

"Yes. I want a check right away on those Nathian records the archaeologists found. Find out if Heleb was one of the planets they seeded."

"Right. Hang on." There was a long silence, then: *"Lew, this is Stet. How come the question about Heleb?"*

"Was it on that Nathian list?"

"Negative. Why'd you ask?"

"Are you sure, Stet? It'd explain a lot of things."

"It's not on the lists, but ... wait a minute." Silence. Then: "Heleb was on line of flight to Auriga, and Auriga was on the list. We've reason to doubt they put anyone down on Auriga. If their ship ran into trouble—"

"That's it!" snapped Orne.

"Keep your voice down or talk subvocally." ordered Stetson. "Now, answer my question: What's up?"

"Something so fantastic it frightens me," said Orne. "Remember that the women who ruled Heleb bred female or male children by controlling the sex of their offspring at conception. The method was unique. In fact, our medics thought it was impossible until—"

"You don't have to remind me of something we want buried and forgotten," interrupted Stetson. "Too much chance for misuse of that formula."

"Yes," said Orne. "But what if your Nathian underground is composed entirely of women bred the same way? What if the Heleb women were just a bunch who got out of hand because they'd lost contact with the main element?"

"Holy Moley!" blurted Stetson. "Do you have evidence—"

"Nothing but a hunch," said Orne. "Do you have a list of the guests who'll be here for the election party tomorrow?"

"We can get it. Why?"

"Check for women who mastermind their husbands in politics. Let me know how many and who."

"Lew, that's not enough to—"

"That's all I can give you for now, but I think I'll have more. Remember that ... " he hesitated, spacing his words as a new thought struck him "... the ... Nathians ... were ... nomads."

* * * * *

Day began early for the Bullones. In spite of its being election day, Bullone took off for his office an hour after dawn. "See what I mean about this job owning you?" he asked Orne.

"We're going to take it easy today, Lew," said Diana. She took his hand as they came up the steps after seeing her father to his limousine flitter. The sky was cloudless.

Orne felt himself liking her hand in his—liking the feel of it too much. He withdrew his hand, stood aside, said: "Lead on."

I've got to watch myself, he thought. *She's too charming.*

"I think a picnic," said Diana. "There's a little lake with grassy banks off to the west. We'll take viewers and a couple of good novels. This'll be a do–nothing day."

Orne hesitated. There might be things going on at the house that he should watch. But no … if he was right about this situation, then Diana could be the weak link. Time was closing in on them, too. By tomorrow the Nathians could have the government completely under control.

It was warm beside the lake. There were purple and orange flowers above the grassy bank. Small creatures flitted and cheeped in the brush and trees. There was a *groomis* in the reeds at the lower end of the lake, and every now and then it honked like an old man clearing his throat.

"When we girls were all at home we used to picnic here every Eight–day," said Diana. She lay on her back on the groundmat they'd spread. Orne sat beside her facing the lake. "We made a raft over there on the other side," she said. She sat up, looked across the lake. "You know, I think pieces of it are still there. See?" She pointed at a jumble of logs. As she gestured, her hand brushed Orne's.

Something like an electric shock passed between them. Without knowing exactly how it happened, Orne found his arms around Diana, their lips pressed together in a lingering kiss. Panic was very close to the surface in Orne. He broke away.

"I didn't plan for that to happen," whispered Diana.

"Nor I," muttered Orne. He shook his head. "Sometimes things can get into an awful mess!"

Diana blinked. "Lew ... don't you ... like me?"

He ignored the monitoring transceiver, spoke his mind. *They'll just think it's part of the act*, he thought. And the thought was bitter.

"Like you?" he asked. "I think I'm in love with you!"

She sighed, leaned against his shoulder. "Then what's wrong? You're not already married. Mother had your service record checked." Diana smiled impishly. "Mother has second sight."

* * * * *

The bitterness was like a sour taste in Orne's mouth. He could see the pattern so clearly. "Di, I ran away from home when I was seventeen," he said.

"I know, darling. Mother's told me all about you."

"You don't understand," he said. "My father died before I was born. He—"

"It must've been very hard on your mother," she said. "Left all alone with her family ... and a new baby on the way."

"They'd known for a long time," said Orne. "My father had *Broach's* disease, and they found out too late. It was already in the central nervous system."

"How horrible," whispered Diana.

Orne's mind felt suddenly like a fish out of water. He found himself grasping at a thought that flopped around just out of reach. "Dad was in politics," he whispered. He felt as though he were living in a dream. His voice stayed low, shocked. "From when I first began to talk, Mother started grooming me to take his place in public life."

"And you didn't like politics," said Diana.

"I hated it!" he growled. "First chance, I ran away. One of my sisters married a young fellow who's now the member for Chargon. I hope he enjoys it!"

"That'd be Maddie," said Diana.

"You know her?" asked Orne. Then he remembered what Stetson had told him, and the thought was chilling.

"Of course I know her," said Diana. "Lew, what's wrong with you?"

"You'd expect me to play the same game, you calling the shots," he said. "Shoot for the top, cut and scramble, claw and dig."

"By tomorrow all that may not be necessary," she said.

Orne heard the sudden hiss of the carrier wave in his neck transceiver, but there was no voice from the monitor.

"What's ... happening ... tomorrow?" he asked.

"The election, silly," she said. "Lew, you're acting very strangely. Are you sure you're feeling all right." She put a hand to his forehead. "Perhaps we'd —"

"Just a minute," said Orne. "About us—" He swallowed.

She withdrew her hand. "I think my parents already suspect. We Bullones are notorious love–at–first–sighters." Her overlarge eyes studied him fondly. "You don't feel feverish, but maybe we'd better—"

"What a dope I am!" snarled Orne. "I just realized that I have to be a Nathian, too."

"You *just* realized?" She stared at him.

There was a hissing gasp in Orne's transceiver.

"The identical patterns in our families," he said. "Even to the houses. And there's the real key. What a dope!" He snapped his fingers. "*The head!* Polly! Your mother's the grand boss woman, isn't she?"

"But, darling ... of course. She—"

"You'd better take me to her and fast!" snapped Orne. He touched the stud at his neck, but Stetson's voice intruded.

"Great work, Lew! We're moving in a special shock force. Can't take any chances with—"

Orne spoke aloud in panic: *"Stet! You get out to the Bullones! And you get there alone! No troops!"*

Diana had jumped to her feet, backed away from him.

"What do you mean?" demanded Stetson.

"I'm saving our stupid necks!" barked Orne. *"Alone! You hear? Or we'll have a worse mess on our hands than any Rim War!"*

There was an extended silence. *"You hear me, Stet?"* demanded Orne.

"O.K., Lew. We're putting the O–force on standby. I'll be at the Bullones' in ten minutes. ComGO will be with me." Pause. *"And you'd better know what you're doing!"*

It was an angry group in a corner of the Bullones' main salon. Louvered shades cut the green glare of a noon sun. In the background there was the hum of air-conditioning and the clatter of roboservants preparing for the night's election party. Stetson leaned against the wall beside a divan, hands jammed deeply into the pockets of his wrinkled, patched fatigues. The wagon tracks furrowed his high forehead. Near Stetson, Admiral Sobat Spencer, the I–A's Commander of Galactic Operations, paced the floor. ComGO was a bull-necked bald man with wide blue eyes, a deceptively mild voice. There was a caged animal look to his pacing—three steps out, three steps back.

Polly Bullone sat on the divan. Her mouth was pulled into a straight line. Her hands were clasped so tightly in her lap that the knuckles showed white. Diana stood beside her mother. Her fists were clenched at her sides. She shivered with fury. Her gaze remained fixed, glaring at Orne.

"O.K., so my stupidity set up this little meeting," snarled Orne. He stood about five paces in front of Polly, hands on hips. The admiral, pacing away at his right, was beginning to wear on his nerves. "But you'd better listen to what I have to say." He glanced at the ComGO. "*All* of you."

Admiral Spencer stopped pacing, glowered at Orne. "I have yet to hear a good reason for not tearing this place apart ... getting to the bottom of this situation."

"You ... traitor, Lewis!" husked Polly.

"I'm inclined to agree with you, Madame," said Spencer. "Only from a different point of view." He glanced at Stetson. "Any word yet on Scottie Bullone?"

"They were going to call me the minute they found him," said Stetson. His voice sounded cautious, brooding.

"You were coming to the party here tonight, weren't you, admiral?" asked Orne.

"What's that have to do with anything?" demanded Spencer.

"Are you prepared to jail your wife and daughters for conspiracy?" asked Orne.

A tight smile played around Polly's lips.

Spencer opened his mouth, closed it soundlessly.

"The Nathians are mostly women," said Orne. "There's evidence that your womenfolk are among them."

The admiral looked like a man who had been kicked in the stomach. "What ...evidence?" he whispered.

"I'll come to that in a moment," said Orne. "Now, note this: the Nathians are mostly women. There were only a few *accidents* and a few planned males, like me. That's why there were no family names to trace—just a tight little female society, all working to positions of power through their men."

Spencer cleared his throat, swallowed. He seemed powerless to take his attention from Orne's mouth.

"My guess," said Orne, "is that about thirty or forty years ago, the conspirators first began breeding a few males, grooming them for really choice top positions. Other Nathian males—the accidents where sex–

control failed—they never learned about the conspiracy. These new ones were full–fledged members. That's what I'd have been if I'd panned out as expected."

Polly glared at him, looked back at her hands.

"That part of the plan was scheduled to come to a head with this election," said Orne. "If they pulled this one off, they could move in more boldly."

"You're in way over your head, boy," growled Polly. "You're too late to do anything about us!"

"We'll see about that!" barked Spencer. He seemed to have regained his self–control. "A little publicity in the right places ... some key arrests and —"

"No," said Orne. "She's right. It's too late for that. It was probably too late a hundred years ago. These dames were too firmly entrenched even then."

* * * * *

Stetson straightened away from the wall, smiled grimly at Orne. He seemed to be understanding a point that the others were missing. Diana still glared at Orne. Polly kept her attention on her hands, the tight smile playing about her lips.

"These women probably control one out of three of the top positions in the League," said Orne. "Maybe more. Think, admiral ... think what would happen if you exposed this thing. There'd be secessions, riots, sub–governments would topple, the central government would be torn by suspicions and battles. What breeds in that atmosphere?" He shook his head. "The Rim War would seem like a picnic!"

"We can't just ignore this!" barked Spencer. He stiffened, glared at Orne.

"We can and we will," said Orne. "No choice."

Polly looked up, studied Orne's face. Diana looked confused.

"Once a Nathian, always a Nathian, eh?" snarled Spencer.

"There's no such thing," said Orne. "Five hundred years' cross–breeding with other races saw to that. There's merely a secret society of astute political scientists." He smiled wryly at Polly, glanced back at Spencer. "Think of your own wife, sir. In all honesty, would you be ComGO today if she hadn't guided your career?"

Spencer's face darkened. He drew in his chin, tried to stare Orne down, failed. Presently, he chuckled wryly.

"Sobie is beginning to come to his senses," said Polly. "You're about through, son."

"Don't underestimate your future son–in–law," said Orne.

"Hah!" barked Diana. "I *hate* you, Lewis Orne!"

"You'll get over that," said Orne mildly.

"Ohhhhhh!" Diana quivered with fury.

"My major point is this," said Orne. "Government is a dubious glory. You pay for your power and wealth by balancing on the sharp edge of the blade. That great amorphous thing out there—the people—has turned and swallowed many governments. The only way you can stay in power is by giving *good* government. Otherwise—sooner on later—your turn comes. I can remember my mother making that point. It's one of the things that stuck with me." He frowned. "My objection to politics is the compromises you have to make to get elected!"

Stetson moved out from the wall. "It's pretty clear," he said. Heads turned toward him. "To stay in power, the Nathians had to give us a fairly good government. On the other hand, if we expose them, we give a bunch of political amateurs—every fanatic and power–hungry demagogue in the galaxy—just the weapon they need to sweep them into office."

"After that: chaos," said Orne. "So we let the Nathians continue … with two minor alterations."

"We alter nothing," said Polly. "It occurs to me, Lewis, that you don't have a leg to stand on. You have me, but you'll get nothing out of me. The rest of the organization can go on without me. You don't dare expose us. We hold the whip hand!"

* * * * *

"The I–A could have ninety per cent of your organization in custody inside of ten days," said Orne.

"You couldn't find them!" snapped Polly.

"How?" asked Stetson.

"Nomads," said Orne. "This house is a glorified tent. Men on the outside, women on the inside. Look for inner courtyard construction. It's instinctive with Nathian blood. Add to that, an inclination for odd musical instruments—the kaithra, the tambour, the oboe—all nomad instruments. Add to that, female dominance of the family—an odd twist on the nomad heritage, but not completely unique. Check for predominance of female offspring. Dig into political background. We'll miss damn few!"

Polly just stared at him, mouth open.

Spencer said: "Things are moving too fast for me. I know just one thing: I'm dedicated to preventing another Rim War. If I have to jail every last one of—"

"An hour after this conspiracy became known, you wouldn't be in a position to jail anyone," said Orne. "The husband of a Nathian! You'd be in jail yourself or more likely dead at the hands of a mob!"

Spencer paled.

"What's your suggestion for compromise?" asked Polly.

"Number one: the I–A gets veto power on any candidate you put up," said Orne. "Number two: you can never hold more than two thirds of the top offices."

"Who in the I–A vetoes our candidates?" asked Polly.

"Admiral Spencer, Stet, myself ... anyone else we deem trustworthy," said Orne.

"You think you're a god or something?" demanded Polly.

"No more than you do," said Orne. "This is what's known as a check and balance system. You cut the pie. We get first choice on which pieces to take."

There was a protracted silence; then Spencer said: "It doesn't seem right just to—"

"No political compromise is ever totally right," said Polly. "You keep patching up things that always have flaws in them. That's how government is." She chuckled, looked up at Orne. "All right, Lewis. We accept." She glanced at Spencer, who shrugged, nodded glumly. Polly looked back at Orne. "Just answer me one question: How'd you know I was boss lady?"

"Easy," said Orne. "The records we found said the ... Nathian (he'd almost said 'traitor') family on Marak was coded as *'The Head.'* Your name, Polly, contains the ancient word *'Poll'* which means *head*."

Polly looked at Stetson. "Is he always that sharp?"

"Every time," said Stetson.

"If you want to go into politics, Lewis," said Polly, "I'd be delighted to—"

"I'm already in politics as far as I want to be," growled Orne. "What I really want is to settle down with Di, catch up on some of the living I've missed."

Diana stiffened. "I never want to see, hear *from* or hear *of* Mr. Lewis Orne ever again!" she said. "That is final, emphatically final!"

Orne's shoulders drooped. He turned away, stumbled, and abruptly collapsed full length on the thick carpets. There was a collective gasp behind him.

Stetson barked: "Call a doctor! They warned me at the hospital he was still hanging on a thin thread!"

There was the sound of Polly's heavy footsteps running toward the hall.

"Lew!" It was Diana's voice. She dropped to her knees beside him, soft hands fumbling at his neck, his head.

"Turn him over and loosen his collar!" snapped Spencer. "Give him air!"

Gently, they turned Orne onto his back. He looked pale, Diana loosed his collar, buried her face against his neck. "Oh, Lew, I'm sorry," she sobbed. "I didn't mean it! Please, Lew ... please don't die! Please!"

Orne opened his eyes, looked up at Spencer and Stetson. There was the sound of Polly's voice talking rapidly on the phone in the hall. He could feel Diana's cheek warm against his neck, the dampness of her tears. Slowly, deliberately, Orne winked at the two men.

THE END

www.ingramcontent.com/pod-product-compliance
Lightning Source LLC
Chambersburg PA
CBHW030045230526
45472CB00005B/1688